CIRCLES AND SPHERES

Sally Morgan

The World of Shapes

Circles and Spheres
Spirals
Squares and Cubes
Triangles and Pyramids

Series editor: Joanna Bentley
Series designer: The Pinpoint Design Company

First published in 1994 by Wayland (Publishers) Ltd, 61 Western Road, Hove, East Sussex,
BN3 1JD, England.

British Library Cataloguing in Publication Data
Morgan, Sally
Circles and Spheres.—(World of
Shapes series)
I. Title II. Series
516

ISBN 0 7502 1285 3

Typeset by Dorchester Typesetting Group Ltd., Dorset, England.
Printed and bound in Italy by Canale, C.S.p.A.

Contents

What is a circle?

A circle is a round shape. It has no straight lines. Every point on the circle is exactly the same distance from the middle.

All the objects in these photographs have a circular shape. This means they are shaped like a circle.

Parts of a circle

The width of a circle is called the diameter. This is the distance from one side of a circle, across the middle, to the other side. The distance from the centre of the circle to the outside is called the radius. The radius is half the diameter.

Measure the diameter and radius of the red circle below.

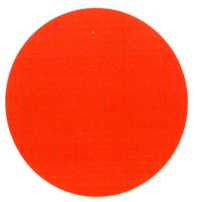

The distance around the outside of a circle is called the circumference. The larger the circle, the bigger the circumference. The children in this photograph are measuring the circumference of this large tree, using a tape measure. As a tree gets older its circumference gets larger.

If you divide a circle into two equal halves you make two semi-circles. You often see semi-circles marked on games pitches.

How do circles fit together?

Circles do not fit closely together. If you tried to push four circles together there would always be a gap, however close you placed the circles.

You can try this for yourself. Take six or seven circular shapes, perhaps coins of the same value. Place them as close together as possible. However you arrange them there will always be a gap. What happens if you use coins of different sizes?

What is a sphere?

A sphere is a solid circle. There are many spheres in your home and school.

We hang colourful spheres on trees at Christmas time. The spheres are called baubles. They come in different colours and are often covered in glitter so they catch the light as they spin.

You probably play games with balls. These are spheres.
A football, a softball and a tennis ball are all spheres.

The blue flowers of this globe thistle are shaped like a sphere. The large flower is really many small flowers packed tightly together to form a sphere.

Piles of spheres

How do spheres pack together? The apples in the photograph have been pushed together. There are gaps between each of the apples.

When you try to pile up spheres lots of spaces appear. It is not easy to make a large pile of spheres. Spheres usually make a triangular shape when they are piled up. This shape is called a pyramid. A pyramid will not fall over very easily.

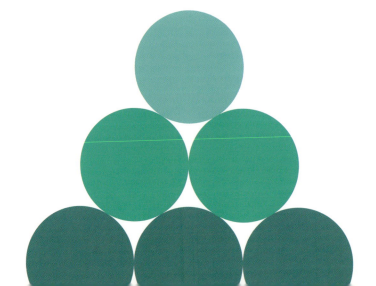

You often see piles of fruit on a market stall. The stall owner piles up the fruit so that it looks attractive and people will want to buy it.

You could try to make a pile of spheres for yourself. You will need at least ten similar sized spheres, such as tennis balls. Which is the most stable way of piling up the spheres?

Circles in the home

If you look around your home or school you will see many different types of circles.

The lids of these containers (right) are all circles. What type of things can you keep in containers?

If you look at the ends of drinks cans you see that they have a circular shape. But if you look at the cans from the side they have a rectangular (oblong) shape. The shape of the drinks can is called a cylinder. ▼

◄ These cotton **reels** are cylinders. The shape is ideal for winding a long length of thread around.

Sticky tape comes in a roll. The ▶
tape is wound around a reel in
the middle. Sticky tape comes in
different thicknesses and
colours. This roll of tape
has a circular hole in
the middle.

This girl is using
a magnifying glass.
A magnifying glass
makes objects seem
larger so that you can see
them more clearly. Magnifying
glasses are usually round. They are very
useful when looking at small objects
such as insects or flowers, or small writing.

Records and compact discs (CDs) are circles. The circles of the records are much larger than those of the CDs. If you have a record and a CD you could measure their diameters. First measure the diameter of a record and then measure the CD. How much larger is the diameter of the record?

These plates and saucers are round. Usually a dinner plate has a much greater diameter than a side plate. A side plate has a larger diameter than a saucer.

This building is circular! It is called an oast house. All the rooms in this building are round. There are no straight walls. It was originally used to dry hops, a type of plant used to make beer. The roofs are the same shape as ice-cream cones.

Edible spheres

Some of the foods you eat are spheres. Next time you are in the kitchen have a look for foods that are shaped like spheres (spherical).

Fruits often have a spherical shape. Although they are all similar in shape, they come in different colours and sizes. Sometimes the fruit is inside a thick skin which has to be peeled first. Have a look at the photograph of the bowl of fruit. Do you know what all the fruits are?

Grapes grow in a **bunch**.
Each grape is a small sphere.
It has a stalk which is joined to the stem to form a bunch. Some grapes are sweet and can be eaten raw. Other grapes are used to make wine.

Some vegetables are spherical as well. There are spherical cauliflowers, cabbages, lettuces, peppers and turnips.

Eggs are not quite spherical. They have an **oval** shape. If you break open an egg and drop it carefully into a frying pan it becomes flat and circular in shape. The centre of the egg is yellow. This part is called the yolk. The rest of the egg is called the egg white. The white is almost see-through to start with but it becomes more solid and white as it cooks.

Many sweets and chocolates are circles or spheres. This shape is easy to pick up and pop in the mouth!

Circles in the street

There are circles everywhere you look in a street.

Many road signs are circular. Circular signs give instructions to the road users.

This sign means that cyclists are **forbidden** to use this path. The circle has a red edge with a drawing of a bicycle in the middle.

This circular sign has a red edge and a thick red line across the middle. It means 'no entry'. Traffic may not enter this road.

The wheels of cars and bicycles are circular. The wheels on this wheelchair have narrow tyres and many **spokes** running from the centre of the wheel to the edge.

This boy is riding a go-cart round a track. How many tyres can you see?

Circles in the countryside

When you visit the countryside, look out for circles.

Many farmers grow fields of grass which they harvest at the end of summer. The grass is cut near the ground and rolled up into **bales**. These bales are circular when seen from the end. The grass is called hay. ▶

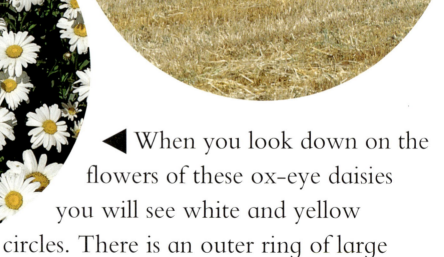

◀ When you look down on the flowers of these ox-eye daisies you will see white and yellow circles. There is an outer ring of large white petals and an inner ring that is yellow. These flowers are very colourful and they attract many insects.

The leaves of the water lily float on the surface of the water. They look like large plates. They are held in place by underwater stems. The leaves are quite strong and small animals, such as frogs, can sit on them. ▼

▲ The sun often looks like a red fiery ball as it sets low in the sky.

All sorts of bubbles

Bubbles are hollow spheres. There is just air inside a bubble. Bubbles of soap are **fragile** and they burst very easily.

Soap bubbles are made when soap is mixed into water. The bubbles form a **froth**. If you look very carefully at a large bubble you can see a reflection of yourself on the surface of the bubble. The bubble is like a mirror.

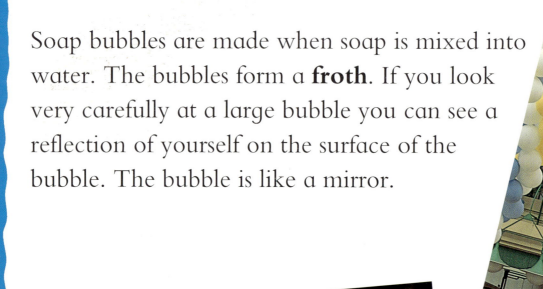

Balloons are hollow too. A balloon is made from coloured rubber. When air is blown into the balloon the rubber expands and becomes spherical in shape. These balloons have been tied up so that they form a ▼ colourful **decoration** in a shopping centre.

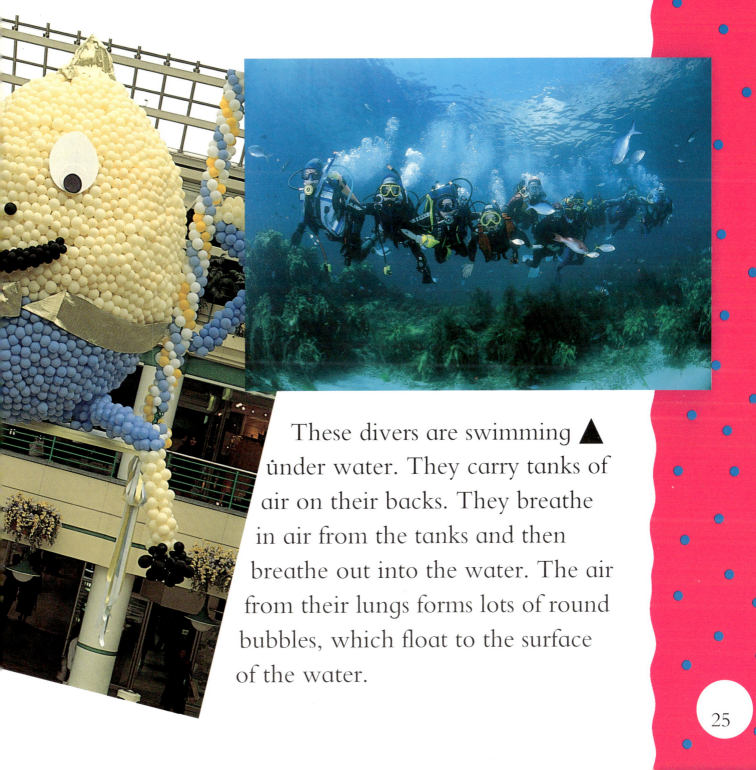

These divers are swimming ▲ under water. They carry tanks of air on their backs. They breathe in air from the tanks and then breathe out into the water. The air from their lungs forms lots of round bubbles, which float to the surface of the water.

Blowing bubbles

Blowing bubbles is easy. All you need is 20 cm of thin wire, water and detergent (such as washing up liquid).

Pour some water into a bowl and add a little detergent. Carefully stir the detergent into the water. Do not make any bubbles. Leave this mix on the side for a few hours. Make a circular shape like a spoon with the wire.

Carefully pull the wire spoon through the water. Lift it out of the water and make sure that you have a film of soapy water across the spoon. Gently blow the water. It will start to bulge and then a bubble will break free.

Make a spinning top

It is easy to make a colourful spinning top with a piece of card and a pencil.

Cut out a circle from stiff white card. You can paint or colour it in patterns like these.

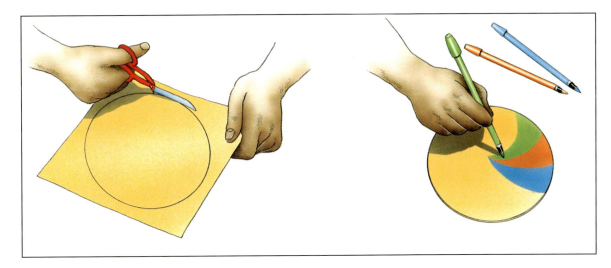

Push a pencil through the middle of the circle, and make the card spin by twisting the top of the pencil.

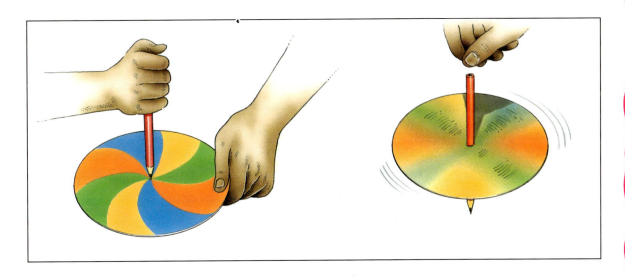

Other shapes

There are many other shapes in this photograph. How many different shapes can you spot?

Look carefully at the pattern of circles below. How many circles can you spot? What is the diameter of the largest circle?

Glossary

Bales Large bundles of hay with rope wound around them.

Bunch A number of things tied together or growing together.

Decoration Something which is attractive.

Forbidden Not allowed.

Fragile Easily broken.

Froth Foam on the top of a liquid.

Oval Shaped like a stretched sphere.

Reel A small piece of wood or plastic on which thread is wound.

Spoke A long thin piece of wood or metal which reaches from the middle to the rim of a wheel.

Tadpole A frog in its early form, before its legs grow.

Books to read

Shapes by Ivan Bulloch (Two-Can, 1994)

Shapes by Jacqueline Dineen (Wayland, 1990)

Circles by Rose Griffiths (A & C Black, 1993)

Shapes by John Satchwell (Walker, 1988)

Index

Picture acknowledgements

The publishers wish to thank the following for providing the pictures for this book: Chapel Studios/Zul Mukhida 5 (top), 16 (left); Ecoscene 6, 11 (right), 14 (both), 15 (top and bottom right), 16 (centre), 17 (right), 18 (both), 19, 20 (both), 21 (both), 22 (both), 23 (both), 24 (right); Eye Ubiquitous 10 (both), 13; Tony Stone Images 4, 10 (left), 12, 24, 25 (right); Wayland Picture Library 5 (bottom), 9, 13 (bottom), 19 (left), 21 (top), 26, 28; ZEFA 7, 15, 24. Artwork by Peter Bull.